Louiza Ouksel

Anti-inflammatory effect of phosphonates

Louiza Ouksel

Anti-inflammatory effect of phosphonates

In Seleco anti-inflammatory test

ScienciaScripts

Imprint

Any brand names and product names mentioned in this book are subject to trademark, brand or patent protection and are trademarks or registered trademarks of their respective holders. The use of brand names, product names, common names, trade names, product descriptions etc. even without a particular marking in this work is in no way to be construed to mean that such names may be regarded as unrestricted in respect of trademark and brand protection legislation and could thus be used by anyone.

Cover image: www.ingimage.com

This book is a translation from the original published under ISBN 978-620-6-71376-0.

Publisher:
Sciencia Scripts
is a trademark of
Dodo Books Indian Ocean Ltd. and OmniScriptum S.R.L publishing group

120 High Road, East Finchley, London, N2 9ED, United Kingdom
Str. Armeneasca 28/1, office 1, Chisinau MD-2012, Republic of Moldova, Europe
Printed at: see last page
ISBN: 978-620-7-68138-9

Anti-inflammatory effect of a phosphonate

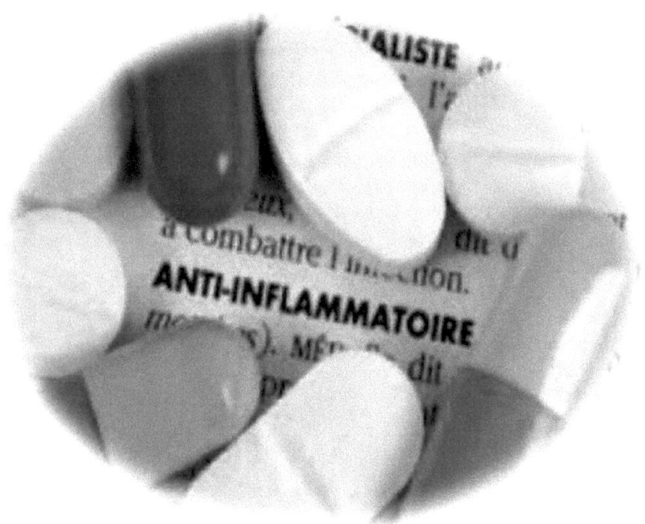

Signing sessions

- ➢ *I* dedicate this work to the memory of my mother, my father and my grandparents, may Allah the Almighty grant them peace and mercy.

- ➢ *To* my husband, who helped me a lot with his advice.

- ➢ *To* my daughters Imen, Amina and Maissa Nour el Houda.

- ➢ *To* my brother Mohammed and my sister Zahia.

FOREWORD

Inflammation is a condition that predates humanity itself, as witnessed by the first signs of inflammatory processes observed on dinosaur bones. Inflammation has always been considered a vital defense mechanism against infection. In the 18th[e] century, John Hunter was one of the first to define inflammation as a beneficial function, a concept that was validated a century later by Elie Metchnikoff. In order to limit the side effects of inflammation, the use of herbal anti-inflammatories was introduced in China as early as 2800 B.C. and in Egypt as early as 1520 B.C.

However, finding an effective new drug with a low damage rate can cost up to US$1.8 billion and take around thirteen years. However, the advent of bioinformatics technologies has made this task easier and cheaper. In our study, we chose a chemical compound synthesized at the Laboratory of Electrochemistry of Materials (LEM) with chemical formula $C12H27O5P$ (DH4MPMP) to perform a bioassay to evaluate its effect as an anti-inflammatory agent.

Chapter I: Is devoted to a bibliographical study

presenting general notions on immunity.

Chapter II: is devoted to a bibliographical study presenting general notions on the inflammatory reaction and the factors that cause inflammation.

Chapter III: presents the properties of phosphonates

Chapter IV: In this chapter we present and discuss the various results obtained by target prediction, molecular docking confirmation, Iipinski rules and ADMET.

Table of contents

Chapter 1: Immunity

1.1 Immunity

Immunity is the ability of living organisms to defend themselves against foreign agents (viruses, parasites, bacteria...). The first line of defense are the natural barriers that separate the organism from its environment: skin and mucous membranes in animals, bark and cell walls in plants. Once past these barriers, foreign agents trigger immune defenses, the complexity of which varies throughout the living world. To protect itself, the human body has two types of defense mechanisms: innate immunity and adaptive immunity.

1.1.1 Innate immunity

This is the body's first line of defense, providing a rapid, non-specific response to invading pathogens or even figurei. It includes physical barriers (the skin), chemical defenses (enzymes, antimicrobial proteins) and different types of immune cells (phagocytes, natural killer cells) that detect and eliminate pathogens.

1.1.2 Adaptive immunity

This is a more specialized and long-lasting defense

mechanism that develops after exposure to specific pathogens. It involves lymphocytes, a type of white blood cell, which undergo a process called clonal selection to generate specific immune responses. This system has the remarkable ability to recognize and remember specific pathogens, enabling a more rapid and targeted response in subsequent encounters - see Figure 1.

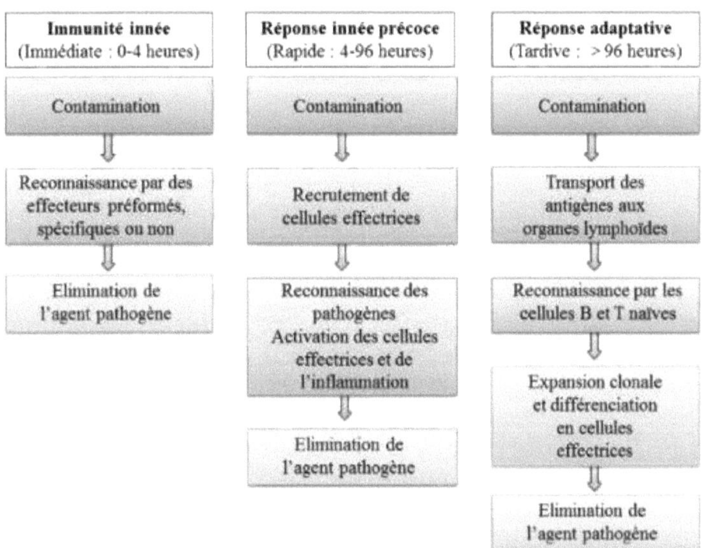

Figure 1: Diagram of the different immune responses.

1.2 Immune system

The immune system is made up of a set of cells and molecules capable, on the one hand, of detecting and recognizing abnormalities and, on the other, of reacting to them. So, when a foreign body enters our body, the immune system can detect it and trigger a series of processes to destroy it. The immune system is regulated by a delicate balance between activation and suppression to prevent dangerous overreactions (autoimmunity) or inadequate responses (immunodeficiency). Disturbances in immune function can lead to a variety of diseases, including autoimmune disorders, allergies and immunodeficiencies.

1.3 Autoimmune diseases

Autoimmune diseases refer to a group of disorders characterized by an abnormal immune response of the body against its own cells, tissues or organs, leading to inflammation and damage. This autoimmune response can affect different parts of the body, leading to a wide range of symptoms and complications. There are over 80 recognized autoimmune diseases, each with its own unique characteristics and target tissues. Some common examples include rheumatoid arthritis, systemic lupus erythematosus, Hashimoto's thyroid, multiple sclerosis, type 1 diabetes and celiac disease. The exact causes of autoimmune diseases are not fully understood, but they probably involve a combination of genetic, environmental and hormonal factors. Certain genes are associated with increased susceptibility to autoimmune disorders, and environmental triggers such as infections, certain drugs and chemical exposures may play a role in initiating or worsening the immune response.

Chapter 2: Inflammation

11.1 Definition of inflammation

Inflammation refers to the body's natural response to injury, infection or irritation. It is a complex biological process involving various cellular and molecular components. The aim of inflammation is to eliminate the initial cause of cellular injury, remove damaged tissue and trigger the healing process. The main features of inflammation include redness, warmth, swelling, pain and loss of function in the affected area. These are collectively known as the cardinal signs of inflammation. However, these 4 signs do not appear every time. Some inflammations do not cause swelling, heat or pain. The inflammation to be most wary of is painless inflammation, with no detectable symptoms. It manifests itself differently, notably through fatigue. Long-term consequences see Figure 2.

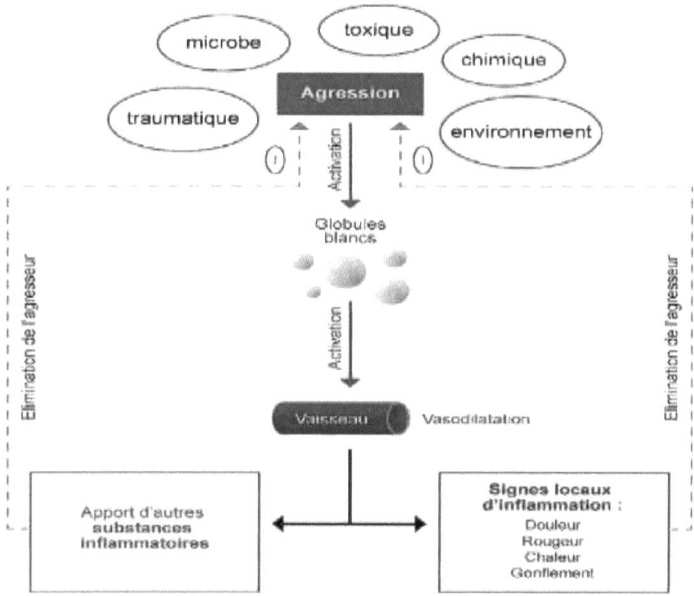

Figure 2: Causes of inflammatory reaction.

II.2 Types of inflammation

Depending on the duration and kinetics of the inflammatory process, inflammation can be classified into two categories:

II.2.1 Acute inflammation

It's a short-term response that usually occurs in response to injury or infection. The main cells involved in acute inflammation are neutrophils, which are white blood cells responsible for attacking and destroying invading pathogens. Acute inflammation is an integral part of the body's defense mechanism and plays a vital role in the healing process.

H.2.2 Chronic inflammation

It is a long-lasting, persistent inflammatory response that can occur when the body fails to eliminate the cause of acute inflammation. It can also occur as a result of autoimmune disorders, chronic infections or prolonged exposure to irritants such as tobacco smoke. Chronic inflammation can lead to tissue damage and is associated with a variety of diseases, including rheumatoid arthritis, inflammatory bowel disease and cardiovascular disease.

II.2.3 Treatment of inflammation

Treatment of inflammation aims to reduce associated symptoms, promote healing and prevent complications. There are different approaches to treating inflammation, depending on its cause and severity. These include:

II.2.3.1 Non-steroidal anti-inflammatory drugs (NSAIDs)

NSAIDs are widely used to reduce inflammation, relieve pain and reduce fever.

They work by inhibiting the production of prostaglandins, which are responsible for inflammation. Common examples include ibuprofen, aspirin and naproxen.

a) Classification of structural NSAIDs

On the basis of their chemical structure, NSAIDs can be classified as salicylates, aryl and heteroaryl acetic acid derivatives, **(Figure 3)** indole/indene acetic acid derivatives, anthranilates and oxicams (enolic acids).

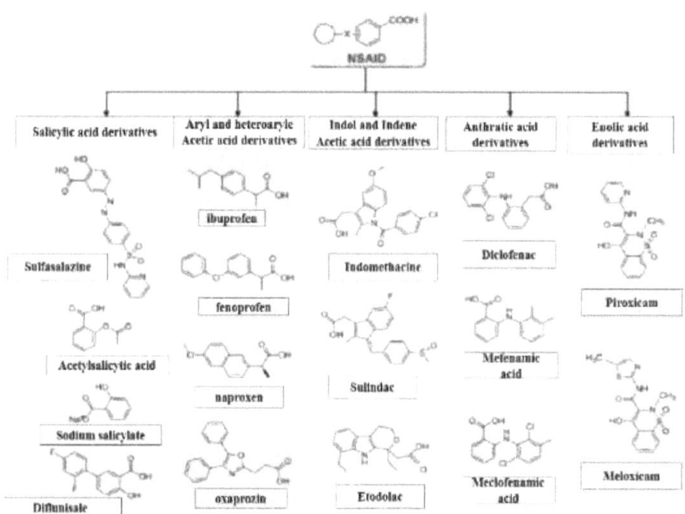

Figure 3: Structural differences between NSAIDs

b) Mechanism of action of non-steroidal anti-inflammatories

The fundamental mechanism of NSAIDs is to inhibit COX enzymes. The two iso forms of the COX enzyme act on the membrane phospholipid known as arachidonic acid to produce various prostaglandins which perform a variety of physiological functions in the body **(Figure 4).**

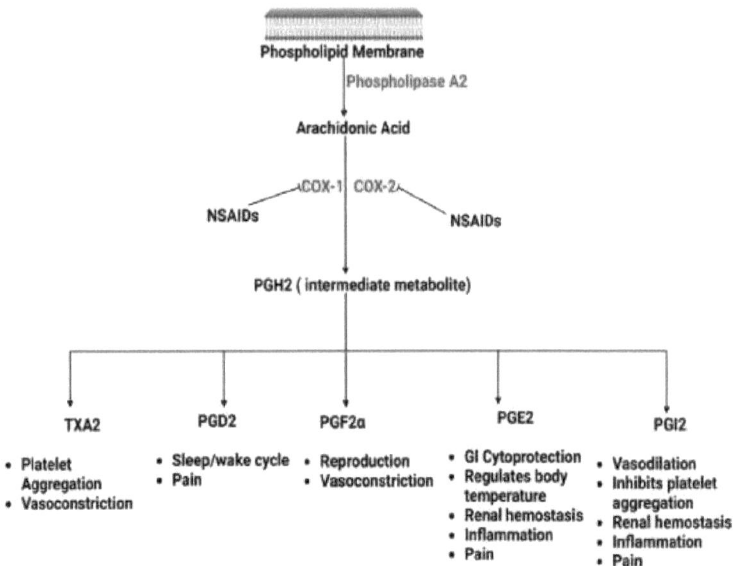

Figure 4: Mechanism of action of NSAIDs.

II.2.3.2 Corticosteroids (anti-inflammatory steroids)

They are synthetic versions of hormones naturally produced by the adrenal glands. Corticosteroids have powerful anti-inflammatory properties and are used to treat a variety of inflammatory conditions. They act by suppressing the immune response and reducing the release of inflammatory substances. Examples include prednisone, hydrocortisone and dexamethasone.

Chapter 3: Phosphonates

111.1   Phosphonate properties

Phosphonates are highly soluble in water, non-volatile and sparingly soluble in organic solvents. They have a threshold effect on the development of salt crystals and a sequestering capacity for metal ions. They are less toxic for the environment and are biodegradable in the soil: release of phosphates. They have particle-dispersing properties and are highly stable under rigorous chemical conditions. They are biologically active compounds. Phosphonates are effective chelating agents that bind strongly to di- and trivalent metal ions, the stability of metal complexes increasing with the number of phosphonate groups.

111.2 Phosphonate applications

Phosphonates are chemical compounds containing a phosphonate group ($-PO_3H_2$) linked to an organic group. These molecules have a particular chemical structure that gives them interesting properties and a variety of applications in different fields.

111.1.1   In biology

Over the past decade, aminophosphonic acids have

been the focus of increased research in the fields of medicine, bio-organic chemistry and organic chemistry, thanks to their diverse biological activities.

111.1.2 In medicine and pharmacology

Nucleoside phosphonates have been extensively studied as potent antiviral agents. Today, more than 40 drugs with a nucleoside structure are used in antiviral chemotherapy. Of these, more than half are used in the treatment of Acquired Immune Deficiency Syndrome (AIDS). Phosphonates are also increasingly used in medicine to treat disorders associated with bone formation and calcium metabolism.

Ribavirin and FdG are the two nucleoside-based compounds with the strongest activity against this type of infection.

III.1.3 Toxicology

Phosphonates are poorly absorbed from the gastrointestinal tract, and most of the absorbed dose is excreted by the kidneys. The toxicity of phosphonates to humans is also low, and they can even be used in the composition of medicines.

Chapter 4: Molecular docking

IV.l Molecular modeling

The first numerical simulations were carried out using simplified models, such as the force-field method, which considered atoms as charged spheres interacting by predefined forces, Over the decades, molecular modeling has benefited from advances in computational techniques and theoretical progress.

Today, thanks to the computing power of modern computers, molecular modeling makes it possible to study complex systems such as proteins, nucleic acids and cell membranes. It is used in many fields of research, such as drug design, materials chemistry and structural biology, contributing to numerous scientific discoveries.

IV. 1.1 Molecular docking

Molecular docking is a computational method that predicts the binding between molecules and their affinity. It is used in drug discovery, particularly in virtual screening. No single program is suitable for all systems, but several powerful programs are available. Molecular docking follows clear steps to achieve good results. Its use in drug discovery is increasing with advances in computing and data availability.

IV. 1.2 Mooring tools

There are several docking tools used for this purpose. Auto-Dock, AutoDock Vina, GOLD, Glide, MOE, ICM and Llex are among the popular software used.

Table 1: Docking software.

Program	Type	Features
MOE	Commercial	Correction of residue problems, structure clean-up, charge assignment based on multiple force fields, protein minimization and binding site prediction.
YASARA	Academic/Commercial	Residue correction, structure cleaning, charge assignment, tautomer assignment, loop remodeling, binding site prediction.
Lead Finder	Commercial	Structure optimization, load allocation, rotamer selection.
BALLView	Academic	Protein minimization and charge allocation.
DeepView	Academic	Protein minimization and load allocation.

AutoDock Tools	Academic	Structure cleaning, charge assignment (Gasteiger), rotamer selection and binding site prediction.
Open Babel	Academic	Structure cleaning, load allocation, Rotamer selection, binding site prediction.

IV. 1.2.1 Molecular Operating Environment (MOE)

MOE (Molecular Operating Environment) is widely used in the pharmaceutical industry and academic research for a variety of drug discovery and molecular modeling applications. Among MOE's key features:

a) Computational chemistry: visualization, manipulation and analysis of molecular structures.

b) Protein modeling: protein structure prediction, homology modeling and protein-ligand docking.

c) Molecular dynamics simulations to study the dynamic behavior of biomolecules.

d) QSAR (quantitative structure-activity relationships) and ADME (absorption, distribution, metabolism and excretion properties) prediction

e) Virtual screening: identify potential drug candidates from large databases of compounds.

f) 3D pharmacophore generation: generate 3D pharmacophore models from protein-ligand complexes or ligand sets.

g) Visualization and analysis: advanced visualization for molecular structures, surfaces and interactions.

IV.   1.2.2 Main methods of molecular docking

Two main methods are used for molecular docking: rigid docking and flexible docking.

a) Rigid docking

This method assumes that the ligand and target molecules retain their fixed three-dimensional structure during binding. The atoms of the molecules are treated as hard spheres, and the calculations focus on the optimal adjustment of the molecules' positions and orientations to maximize complementarity.

b) Flexible docking

In contrast to rigid docking, this method takes into account the possible movements and deformations of the molecules during binding. Flexible regions of the target and/or ligand are allowed to change conformation to optimize fit and improve affinity.

IV.1.2.3 Molecular docking stages

In molecular docking studies, preparation of the target protein and ligand, selection of appropriate docking methods and evaluation of docking results are crucial steps. First, the structure of the target protein is retrieved from a database, its quality checked, and the protein prepared for docking by removing water molecules, ligands and ions. Next, the ligand is prepared by defining the atoms to be included in the template and assigning atomic charges. The ligand can be obtained from a database or designed interactively. Next, the type of docking to be used, such as rigid or flexible docking, is determined according to the characteristics of the target protein and the ligand. In addition, the selection of an appropriate docking scoring function is essential, as it enables the quality of the generated docking patterns to be assessed by evaluating the complementarity between the ligand and the target protein.

Finally, validation of the molecular docking results is achieved by comparing the predicted results with known experimental structures of ligand-protein complexes, or by using experimental techniques such as X-ray crystallography. These steps are fundamental to successful molecular docking studies and contribute to our understanding of protein-ligand interactions, providing valuable information for drug discovery and design.

To perform the molecular docking, we used Molecular Operating Environment (MOE), a molecular modeling program developed by Chemical Computing Group Inc. in Canada. In addition, for the construction of the molecule, we used other software such as ChemDraw and GaussView. The molecule was also optimized using Gaussian9 software. These tools enabled us to take a comprehensive approach to molecular modeling and to carry out docking studies as well as efficiently optimizing the molecule.

IV.1.2.4. Target selection

A therapeutic target is any element of an organism to which an entity modifying its behavior primarily binds, such as an endogenous ligand, drug or medication. Examples of common therapeutic targets are proteins and nucleic acids.

Biological activity refers to the effect of a substance, such as a molecule or drug, on a specific biological system. The results of biological activity evaluation provide information on the effectiveness of the substance in producing the desired effect on the target biological system. These results are important for the selection of promising compounds and the optimization of drugs under development, but require further studies for validation and confirmation.

In order to select therapeutic targets for our ligand, we used the SwissTarget software. The results obtained are illustrated in figure 4 and summarized in table III.l. Among the targets given by the software, we chose three: JAK2, JAK3 and CXCRl to test the anti-inflammatory effect of our synthesized molecule.

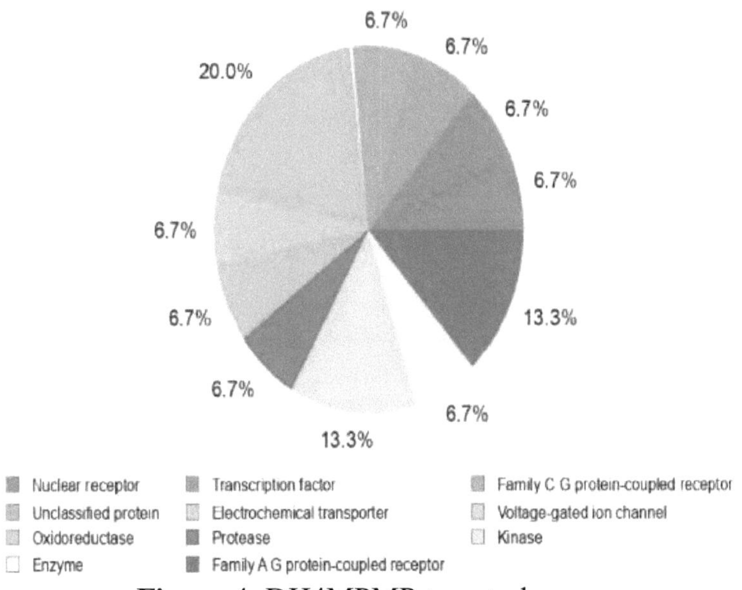

Figure 4: DH4MPMP target classes.

Table 2: Bioactivity of DH4MPMP.

Targets	Name	ID	Class

Tyrosine-protein	Commune JAK2	Uninrot P52333	Kinase
Tyrosine-protein	JAK3	060674	Kinase
Interleukin-8	CXCRl	P25024	RCPG

IV. 2 Docking DH4MPMP

In this section, we present and discuss the docking results of DH4MPMP with the JAK2, JAK3 and CXCRl proteins using both 3D and 2D representations. Full optimization of the DH4MPMP ligand geometry (Figure III. 3) was performed by means of the density functional theory (DFT)-based Gaussian program 09 W, using Beck's three-parameter hybrid functional exchange with 6-31G (d. p) base sets and a functional Lee-yang-Parr correlation (B3LYP).

IV.2.1 The case of JAK2

During docking calculations, we determined 10 poses. Here we present the best pose (Figure III. 4), corresponding to the minimum energies with an RMSD of 1.41Å of the ligand in the JAK2 protein (Table III.10). Figure 5 shows the 2D interactions between the DH4MPMP ligand and the JAK2 protein, noting the presence of H- acceptor interactions with several residues such as Asn673, Asn678 and Thr 555 **(Figure 5).**

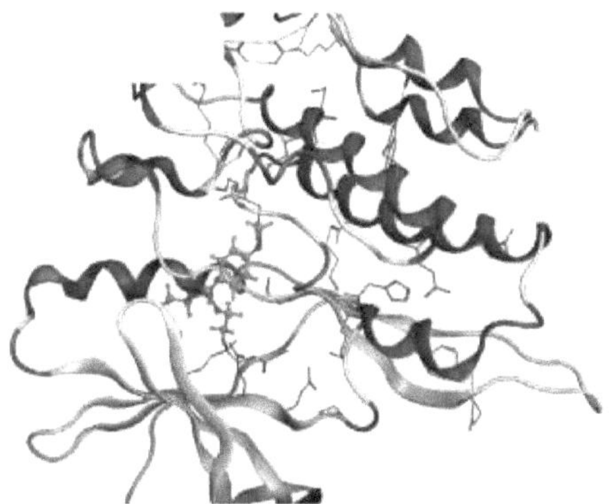

Figure 5: Best 3D position of DH4MPMP in the JAK2 protein.

IV.2.2 The case of JAK3

Figure 6 shows the best pose resulting from the docking of our ligand (DH4MPMP) with the JAK3 protein. This pose corresponds to the minimum energies with an RMSD equal to 1,20Â (Table 3). Three types of interactions were identified: the H-Л interaction with residue Phe833, H-Donor with residue Glu871 and H-

Acceptor with Lys855 residue **(Figure 6)**

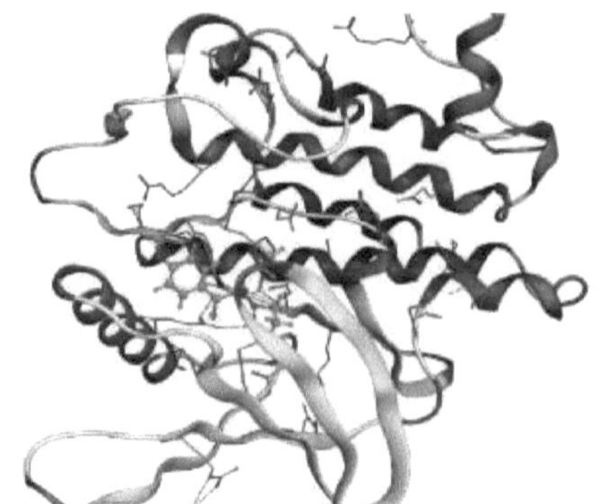

Figure 6: Best 3D position of DH4MPMP in the JAK3 protein.

IV.2.3 The case of CXCRl

The best pose obtained after docking of our ligand (DH4MPMP) with the CXCRl protein is illustrated in Figure 7, we chose this pose using its RMSD which is 1.1OÂ (Table 3). Two types of interaction were identified: H-π interaction with residue phe211 and π-π interaction with residue phe251 **(Figure 7).**

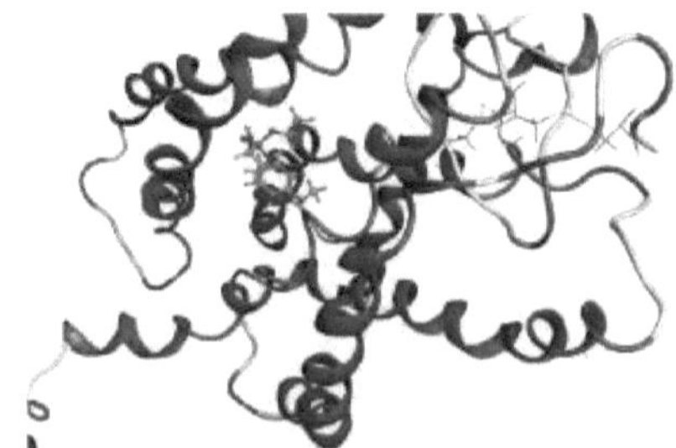

Figure 7: Best 3D position of DH4MPMP in the CXCRl protein.

IV.3 Docking of study protein inhibitors

Comparing the docking results of our ligand with those of the JAK2, JAK3 and CXCR1 inhibitors enables us to analyze several important properties, such as compliance with Lipinski's rule, energies, interactions obtained by molecular docking and pharmacokinetic properties.

IV.3.1 The case of JH2

The best pose obtained after docking of the JH2 inhibitor with the JAK2 protein is illustrated in figure 8, and we chose this pose using its RMSD of 1.12Å (Table 3). The interaction between JAK2 and JH2 shows spatial proximity at the binding site. The JH2 inhibitor forms an H-acceptor bond with the tyrosine amino acid, where the O group of JH2 approaches the tyrosine amino acid. In addition, another interaction occurs between the O group of JH2 and the lysine acid, where the bond is of the H-acceptor type. These interactions are crucial for stabilizing the interaction between JAK2 and JH2.

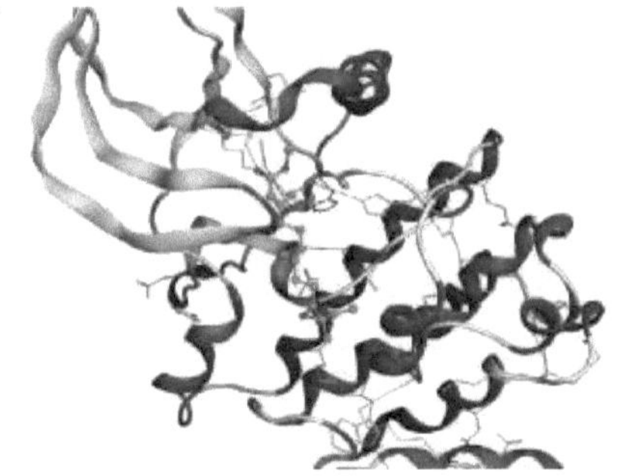

Figure 8: Best 3D position of the JH2 inhibitor in the JAK2 protein.

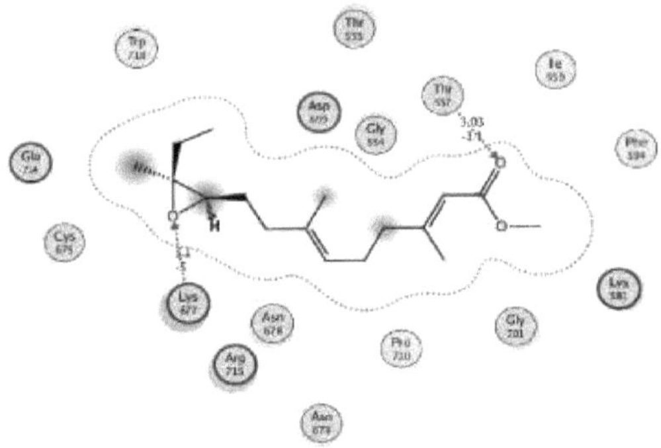

Figure 9: 2D representation of interactions between the JAK2 protein and the JH2 inhibitor.

IV.3.2 Leflunomide

The best pose obtained after docking of the Leflunomide inhibitor with the JAK3 protein is illustrated in figure III.12, and we chose this pose using its RMSD of 1.55Â (Table 3). The inhibitor Leflunomide forms an H-donor bond with the amino acid Asp967, where the N6 group of

Leflunomide approaches this residue. A similar interaction occurs between Leflunomide's C28 group and the same residue. The Leflunomide ring exhibits a π - Cation interaction with residue Lys855 **(Figure 10).**

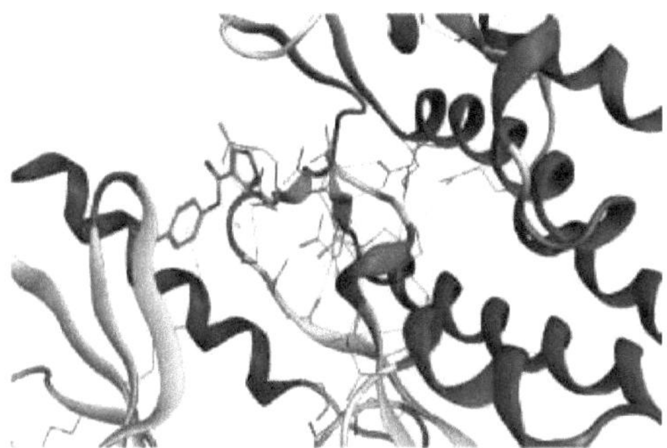

Figure 10: Best 3D position of the Leflunomide inhibitor in the JAK3 protein.

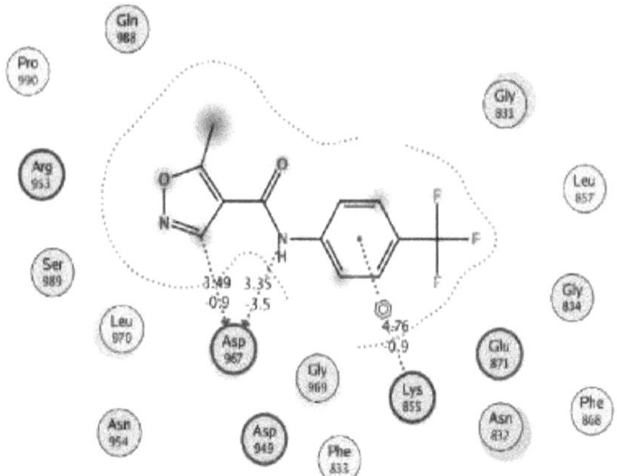

Figure 11: Best interaction position of JAK3 with Leflunomide in 2D.

IV.3.3 Ladarixine

The best pose obtained after docking of the **Ladarixin** inhibitor with the JAK3 protein is illustrated in figure III.14, and we chose this pose using its RMSD of 1.76Â (Table IILlO). The **Ladarixin** inhibitor forms an H- acceptor bond with the Glu291 residue, where the C22 group of the **inhibitor** approaches this residue. In addition, another π -H interaction occurs between the same group and residue Phe251 **(Figure 12).**

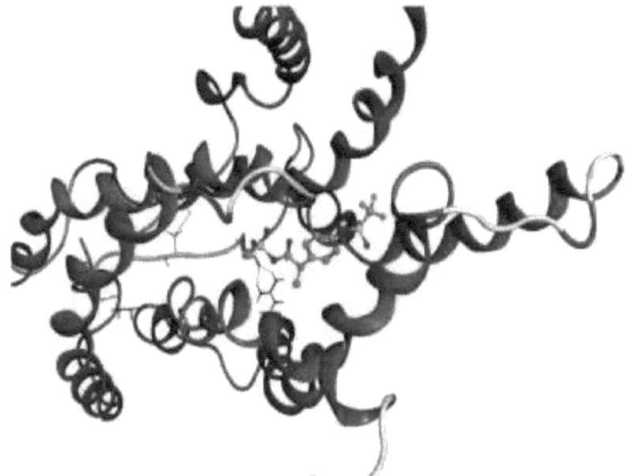

Figure 12: Best 3D position of the **Ladarixin** inhibitor in the CXCRl protein.

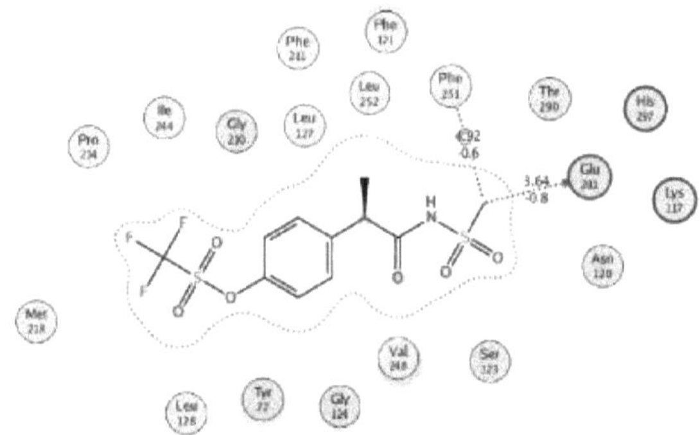

Figure 13: Best interaction position of CXCR1 with Ladarixine.

Analysis of the results of DH4MPMP docking in the CXCR protein gave a docking score of -6.50 Kcal/mol, almost comparable to that of the drug Ladarixine, which is equal to -6.67 Kcal/mol, so we can say that DH4MPMP may be a drug yet to be proven experimentally (see **Table 3).**

Table 3: Molecular docking results and interactions.

Protein	Ligand	Score Kcal/mol	RMSD	Interactions		
				Atom	Residues	Connection type
JAK2	1	−5.74	1.41	O_{18}	Asn673	H- acceptor
				O_{18}	Asn678	H- acceptor
				O_{36}	Thr555	H- acceptor
	2	−7.40	1.12	O_4	Thr557	H- acceptor
				O_6	Lys677	H- acceptor

JAK3	1	⁻6.25	1.20	Cycle C_{11} O_{36}	Phe833 Glu871 Lys855	H-π H-Donor H-Acceptor
	3	⁻5.20	1.55	Cycle N_6 C_{18}	Lys855 Asp967 Asp967	π-Cation H-Donor H-Donor
CXCR1	1	⁻6,50	1.10	O_{34} Cycle	Phe211 Phe251	H-π π - π
	4	⁻6,67	1.76	C_{22} C_{22}	Phe251 Glu291	H-π H-Acceptor

(1) DH4MPMP, (2) JH2, (3) Leflunomide, (4) Ladarixine.

IV.4 Lipinski's rule

Using Lipinski's rule as a reference, which is commonly used to assess the "drug- likeness" of compounds, we compare the values obtained for each compound (Table 3).

Starting with molecular weight, the compounds respect Lipinski's rule, as their molecular weights are below 500 g/mol. This indicates that they have an appropriate molecular size to be considered as potentially pharmacologically favorable compounds.

As far as hydrogen bond acceptors (HBAs) are concerned, the compounds also respect Lipinski's rule, as they have fewer than 10 HBA groups. This suggests that they may be capable of establishing appropriate interactions with other molecules in a biological context.

However, when it comes to hydrogen bond donors (HBDs), this can influence the compound's ability to form specific bonds in a biological environment.

With regard to the octanol-water partition coefficient (LogP), the compounds comply with Lipinski's rule, as they have a LogP of less than 5. This value suggests adequate solubility in both hydrophobic and hydrophilic solvents.

For total polar surface area (TPSA), the compounds respect Lipinski's rule, as their TPSA is less than 140 $Å^2$. This indicates a moderate total polar surface area, which may be favorable from a pharmacological point of view.

We can say that the compounds meet most of the criteria of Lipinski's rule. However, it's important to note that Lipinski's rule is a rule of thumb, and there are many other factors to consider when assessing a compound's viability and efficacy as a potential drug.

From these results we can say that the DH4MPMP ligand could be used as a drug in comparison with the inhibitors of the three proteins that are already marketed.

(Table 4)

Table 4: Lipinski's Rule analysis for DH4MPMP and the three inhibitors.

Formula	Weight (g/mol)	HBA	HBD	LogP	TPSA ($\hat{A}$)^2
$C_{12}H_{19}O_5P$	274.25	5	1	2.79	74.80
$C_{17}H_{28}O_3$	280.40	3	0	4.18	38.83
$C_{12}H_9F_3N_2O_2$	270.21	3	1	3.25	55.13
$C_{11}H_{12}F_3NO_6S_2$	375.34	9	1	4.52	123.37
Lipinski's rule	≤500	<10	<5	<5	<140

IV.5 DH4MPMP ADMET

The results of ADMET evaluation are essential for selecting the most promising molecules for the subsequent development phase and helping to avoid costly failures. They also help optimize drug design by providing information on the modifications needed to improve ADMET properties. It should be noted that the results of ADMET evaluation are not definitive and may vary depending on a variety of factors. Further experimental studies, such as in vitro and in vivo assays, are required to validate and refine these results to ensure the safety and efficacy of drugs under development.

Table 5 shows the ADMET results for the DH4MPMP ligand. The DH4MPMP ligand exhibits human intestinal absorption with a probability of 0.94, indicating good absorption through the intestinal wall. It is able to cross the blood-brain barrier with a probability of 0.72, meaning it can enter the brain from the bloodstream. It does not present respiratory, hepatic or renal toxicity. **(Table 5)**

Table 5: ADMET analysis for DH4MPMP.

Classification	Value	Probability
Human intestinal absorption	+	0,9393
Blood-brain barrier	+	0,7250
Carcinogenicity (binary)		0,7428
Eye corrosion		0,8129
Hepatotoxicity		0,6734
Skin sensitization		0,8515
Respiratory toxicity		0,5889
Nephrotoxicity		0,5315
Connection to the	+	0,5771
Connection to the	+	0,5426
Connection to the		0,5562
Biodegradation		0,7500

IV.6 Biological activity of DH4MPMP

Biological activity refers to the effect of a substance, such as a molecule or drug, on a specific biological system. The results of the biological activity evaluation provide information on the effectiveness of the substance in producing the desired effect on the target biological system. The results of the biological activity evaluation provide information on the efficacy of a substance in producing the desired effect on a specific biological system. These results are important for the selection of promising compounds and the optimization of drugs under development, but require further studies for validation and confirmation.

Table 6: Biological activity analysis of DH4MPMP.

Score	Score Standardized	Substance
0,912	0,006	Aspulvinone dimethylallyl transferase inhibitor
0,890	0,007	Nootropic
0,847	0,005	Antiarthritic
0,786	0,004	Treatment of atherosclerosis

IV.7 Conclusion

The results of the molecular docking of the ligand alone and with the protein inhibitors used showed us the interactions of this ligand with different residues and with several types of binding.

Using Lipinski's rule as a reference, we can say that the DH4MPMP ligand could be used as a drug in comparison with the inhibitors of the three proteins that are already on the market.

The results of the ADMET evaluation indicate that the DH4MPMP ligand exhibits human intestinal absorption with a probability of 0.94, indicating good absorption through the intestinal wall. It is able to cross the blood-brain barrier with a probability of 0.72, meaning it can enter the brain from the bloodstream. It does not present respiratory, hepatic or renal toxicity.

References

- Coumia, Z., Allen, B., & Sherman, W. (2017). Relative Binding Free Energy Calculations in Drug Discovery: Recent Advances and Practical Considerations. Journal of chemical information and modeling, 57(12), 2911-2937...
- Ouksel, L., Chafaa, S., Bourzami, R., Hamdouni, N., Sebaisb, M., & Chafai, N. (2017). Crystal structure, vibrational, spectral investigation, quantum chemical DFT calculations and thermal behavior of Diethyl [hydroxy(phenyl) methyl] phosphonate.
- M. Mehri, N. Chafai, L. Ouksel, K. Benbouguerra, A. Hellal, S. Chafaa, Synthesis, electrochemical and classical evaluation of the antioxidant activity of three α-aminophosphonic acids: Experimental and theoretical investigation, J. Mol. Struct. 1171 (2018)179-189.
- Cleyrat, C., Jelinek, J., Girodon, F., Boissinot, M., Ponge, T., Harousseau, J. L., Issa, J. P., & Hermouet, S. (2010). JAK2 mutation and disease phenotype: a double L611V/V617F in cis mutation of JAK2 is associated with isolated erythrocytosis and increased activation of AKT and ERK1/2 rather than STAT5. Leukemia, 24(5), 1069-1073.
- Schindler C. W. (2002). Series introduction. JAK-STAT signaling in human disease. The Journal of clinical investigation, 109(9), 1133-1137.
- L. Ouksel, R. Bourzami, S. Chafaa, N. Chafai, Solvent and catalyst-free synthesis, corrosion protection, thermodynamic, MDS and DFT calculation of two en Vironmentally friendly inhibitors: Bis-phosphonic

acids, J. Mol. Struct. 1222 (2020) 128813.
- Sitaru, S., Budke, A., Bertini, R., & Sperandio, M. (2023). Therapeutic inhibition of CXCR1/2: where do we stand? Internal and emergency medicine, 1-18. Advance online publication.
- Lipinski C. A. (2000). Drug-Iike properties and the causes of poor solubility and poor permeability. Journal of pharmacological and toxicological methods, 44(1), 235-249.
- Bessout, R. (2012). Treatment of radiation-induced colorectal lesions by Mesenchymal Stromal Cell (MSC) injection: involvement of the inflammatory process.
- website list
    http://mas.stephanie.free .fr/cours systeme im munitaire.htm
    https://www.aarda.org/
    https://public.larhumatologie.fr/linflammation
    https://celluloyd.tumblr.com/post/163341684741/r%C3%A9cepteurs-coupl%C3%A9s-%C3%A0-nf-%CE%BAb-en-communication

I want morebooks!

Buy your books fast and straightforward online - at one of world's fastest growing online book stores! Environmentally sound due to Print-on-Demand technologies.

Buy your books online at
www.morebooks.shop

Kaufen Sie Ihre Bücher schnell und unkompliziert online – auf einer der am schnellsten wachsenden Buchhandelsplattformen weltweit! Dank Print-On-Demand umwelt- und ressourcenschonend produziert.

Bücher schneller online kaufen
www.morebooks.shop

Printed by Books on Demand GmbH, Norderstedt / Germany